My Sticker Activity Book

Diggers and Dump Trucks

How To Use This Book

Each double page has an introduction to Diggers and Dumpers and lots of shaded areas where you can put your stickers. You'll find the stickers on sheets in the middle of this book. To help you find the correct page for your stickers, each sticker has a coloured track mark beside it. Find the matching colour spaces in the book. The stickers are also numbered to help you position them on the page. There are lots of extra stickers and fun activities to do, including quizzes, wordsearches and colouring. The answers are on page 12.

Written by Patricia Moss • Illustrated by Simon Abbott • Picture research by Hannah Anderson

The publishers would like to thank the following for their kind permission to reproduce photographs:
Volvo Construction Equipment Ltd • Caterpillar • Case IH • Liebherr Great Britain • Case Construction Equipment, a division of CNH Global NV • Dynapac • Thwaites Dumpers Ltd • DaimlerChrysler UK Limited • Chris Hodge Commercial, London, www.chrishodgetrucks.com • Parker Plant Ltd • Peter Breitbach/SCANIA • Jonas Nordin/SCANIA • Dan Boman/SCANIA • Conny Hetting/SCANIA • Ingemar Eriksson/SCANIA • Komatsu Ltd

Diggers and dumpers

Giant construction machines, such as diggers and dumpers, are found on all building sites. Diggers have special scoop-like attachments that can lift up huge mountains of soil, or buckets with metal teeth that can dig trenches in the ground. Dumpers transport tonnes of material and empty it out wherever it's needed.

This dump truck is unloading sand at a quarry.

Mini excavators are used in small spaces to dig trenches and shift earth.

Wheel loaders can shovel, lift, dig and move earth.

Track excavators have caterpillar tracks to keep them stable on uneven ground.

Giant wheel loaders are used in quarries.

Amazing fact

The largest excavators can weigh as much as 25 fully grown elephants.

This dump truck has an extra powerful engine to pull two trailers.

This backhoe loader has a scoop on the front and a digging arm on the back.

Find the words listed below in the word search.

E	E	I	P	L	A	Y	M	D	D
O	X	J	U	R	R	X	A	O	O
E	B	C	P	R	A	S	C	D	G
S	L	A	A	W	M	F	H	D	S
N	H	U	C	V	T	E	I	U	R
T	Q	O	E	K	A	Y	N	M	E
R	E	I	V	L	H	T	E	P	G
U	O	X	L	E	T	O	O	E	G
C	I	T	C	H	L	S	E	R	I
K	I	T	S	O	I	L	O	R	D

machine

quarry

backhoe

digger

truck

shovel **dumper**

excavator **soil**

The container at the back of a dump truck is called a hopper.

A large piston lifts up the hopper, and the load slides off the back.

This dump truck has extra large wheels to spread the weight of its load.

This cab can pull two tipping trailers.

Special attachments

Diggers can be fitted with all kinds of attachments to turn them into grabbers and loaders. Some are even fitted with pneumatic corkscrew drill attachments which can break through tough ground. Variations of these attachments are fitted to cranes and trucks.

This corkscrew drill attachment is breaking through soil.

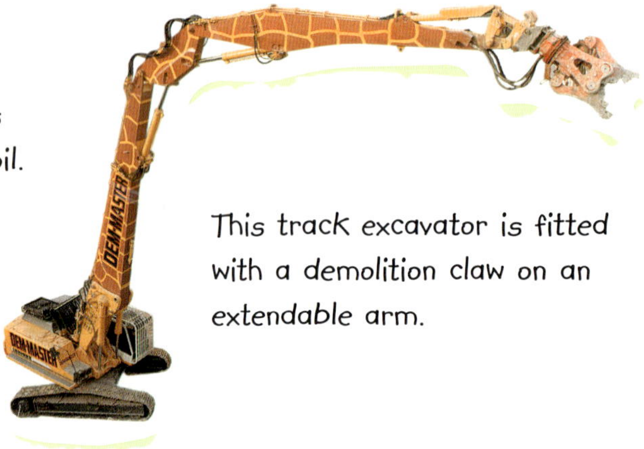

This track excavator is fitted with a demolition claw on an extendable arm.

This dumper truck is fitted with a claw on a hinged arm for lifting logs.

Unscramble the words below.
You'll find the words on these two pages.

tidmoonlie lawc

_ _ _ _ _ _ _ _ _ _   _ _ _ _

lirkfoft

_ _ _ _ _ _ _ _

pirepgr

_ _ _ _ _ _ _

This tractor is fitted with a gripper attachment for transporting logs.

Amazing fact

The driver's cab of demolition machines is fitted with bullet proof glass.

The huge claw attachment on this track excavator can pull down buildings.

This tractor has stabilisers to steady it when the fork attachment is fully extended.

This tooth-edged digger bucket can scrape earth from the ground.

This tractor has a forklift attachment on an extendable arm to lift and move things.

This crane attachment has a strong clip to lift material.

This pneumatic drill attachment can break up solid concrete.

Where to find them

Most kinds of heavy machines are found on building sites. They are tough enough for working on uneven ground and can cope with lifting and carrying heavy materials. Diggers and dumpers are also used on scrap metal yards, road construction sites, quarries, mines and demolition sites.

This track excavator is lifting a concrete block on a building site.

This excavator is scooping rubble up in its bucket to load into a dump truck.

This dumper truck is tipping its load of sand from a quarry.

Answer the following questions.

1. List 4 places where you can find diggers and dumpers ....

....

....

2. Which machine is used to dig huge foundation trenches? ....

....

3. What is a paver used for? ....

....

This cement mixer is delivering concrete to a house-building site.

5. Giant wheeloaders are used in quarries

1. Backhoe loader breaking up road

5. Giant dump truck tipping its load

2. Track excavator with demolition claw

7. Grader smoothing soil

2. Diggers and dumpers work together

This scraper carries up to 25 tonnes of soil

9. Paver being loaded with small stones

2. Excavator scooping rubble

7. Hinged scoop on backhoe loader

2. Mini excavators dig trenches

4. Backhoe loader transporting pipes

3. Dumper truck fitted with a claw

5. This tractor has stabilisers

1. This track excavator is lifting a concrete block

1. This dump truck is unloading sand

Rollers smooth and flatten

2. Pneumatic drill attachment

5. Bulldozer shifting rubble

Bulldozer with huge blade

9. Crane attachment with metal clip

Double-tipper

Enormous tankers carry heavy loads

7. Hopper on back of dump truck.

3. Dumper truck tipping its load

4. Cement mixer delivering concrete

8. Pneumatic drill attachment

3. Giant recovery truck

7. Tractor with double shovel attachment

9. Dump truck with extra large wheels

8. Digging bucket shifting earth

9. Road paver lays final layer of a road

6. Graders make flat surfaces

6. Track excavator loading up a dump truck

4. The cab of this loader is fitted with extra thick glass

8. This giant dump truck is 5 metres tall

8. Cement mixer

6. Tooth-edged digger bucket

8. Dump truck unloading sand

10. Cab pulling two tipping trailers

7. Excavator fitted with a drill

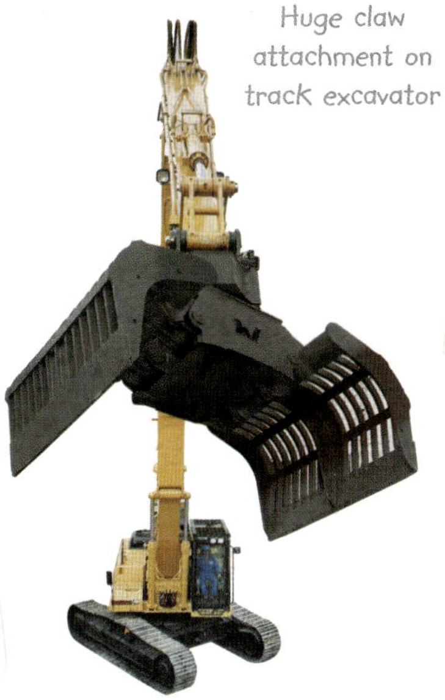

Huge claw attachment on track excavator

6. Backhoe loader with scoop on front and digger on back

Dredger digging mud

1. This backhoe loader can scoop up three tonnes of rock

1. Cork screw drill attachment

3. Cold planer removes top surface

5. Two trailers used to remove gravel

3. Wheel loaders shift, lift and dig

6. This wheel excavator digs foundation trenches

4. Tractor with gripper attachment

4. Track excavators have caterpillar tracks

Amazing fact

Some dump trucks are as large as a house!

This giant dump truck is tipping its load on a building site.

This wheel excavator digs huge foundation trenches for new buildings.

This excavator has been fitted with a drill to break up an old road.

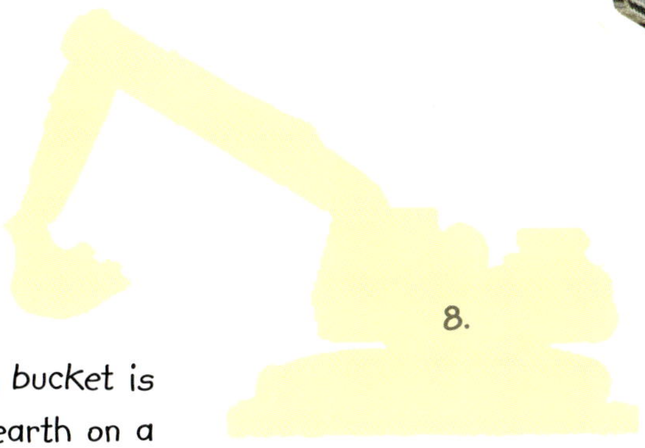

This digging bucket is shifting earth on a road construction site.

This is a paver. It is being loaded up with small stones to make a new road surface.

Not all diggers work on land. This dredger digs up mud from the bottom of the sea.

Heavy loads

We rely on powerful machines to shift heavy loads. On building sites, mounds of earth are dug up for foundations. At quarries, rocks and sand are excavated for their minerals and on demolition sites, old buildings are knocked down. All of this material is loaded up and removed by diggers and dumpers.

This backhoe loader can scoop up nearly 3 tonnes of rock.

Diggers and dumpers work together to remove rubble from building sites.

Giant recovery trucks are used to tow away machines that have broken down on building sites.

Complete the sentences below by filling in the missing words.

1. A __________ loader can scoop up nearly three tonnes of rock.

2. In order to push great loads of earth aside a ____________ is fitted with a huge ______.

3. The extra thick glass of a loader keeps the ___ quiet for the _______.

This bulldozer is fitted with a huge blade at the front to push great loads of earth to one side.

Amazing fact

Bulldozers can work underwater. They create ditches to lay pipes in.

The cab of this loader is fitted with extra thick glass to help keep the cab quiet for the driver.

This huge track excavator is loading a giant dump truck with sand at a quarry.

At large quarries, cabs with two trailers can be used to remove gravel.

The scoop on this backhoe loader is hinged so its load can be easily dropped into a waiting truck.

Some giant dump trucks are five metres tall.

Making roads

The scoop of this backhoe loader is breaking up an old road surface.

Many different machines are used to make roads. Backhoe loaders break up fresh ground and old road surfaces. The rubble is cleared by excavators and loaded into dumper trucks. Graders flatten the ground and cement mixers pour concrete on the new surface. Finally, road pavers lay the top surface and giant rollers flatten the new road surface.

This is a pneumatic drill which is used to prepare a new road.

The sentences below describe how to make a road but they have been mixed up. Write the letters which are next to the sentences in the correct order in the boxes.

1. ☐ 2. ☐ 3. ☐ 4. ☐ 5. ☐ 6. ☐

a. Backhoe loaders break up the old road surfaces.
b. The rubble is cleared by the excavator.
c. Road pavers lay the top surface.
d. Graders flatten the ground.
e. Giant rollers flatten out the new road surface.
f. Cement mixers pour concrete on the new surface.

3.

If only the surface of a road needs repairing, a cold planer is used to remove the top layer.

This backhoe loader is transporting new pipes to lay under a road.

4.

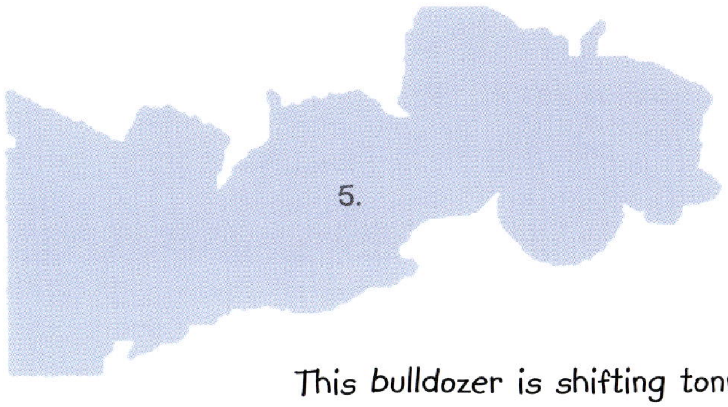

This bulldozer is shifting tonnes of rubble from an old road surface.

Graders have blades to push the soil to one side to make a flat, even surface.

Graders are also used to create smooth slopes.

This road paver lays the final layer of a road. It spreads a thin layer of sticky black asphalt and stones.

Cement mixers prepare and pour concrete to create new road surfaces.

Rollers, like this one, are used to smooth and flatten the top layer of a new road.

Amazing fact

The gigantic wheels on rollers are filled with water to make them extra heavy.

Colouring fun

Colour in the scene below.

Answers

Page 3:

E	E	I	P	L	A	Y	M	D	D
O	X	J	U	R	R	X	A	O	O
E	B	C	P	R	A	S	C	D	G
S	L	A	A	W	M	F	H	D	S
N	H	U	C	V	T	E	I	U	R
T	Q	O	E	K	A	Y	N	M	E
R	E	I	V	L	H	T	E	P	G
U	O	X	L	E	T	O	O	E	G
C	I	T	C	H	L	S	E	R	I
K	I	T	S	O	I	L	O	R	D

Page 4:

demolition claw

forklift

gripper

Page 6:

1. You can choose from the following: building sites, scrap metal yards, road construction sites, quarries, mines and demolition sites
2. wheel excavator
3. To make new road surfaces.

Page 8:

1. backhoe
2. bulldozer/blade
3. cab/driver

Page 10:

The correct order is:

a, b, d, f, c, e